DESIGN IN THE HIGH STR

DESIGN IN THE HIGH STREET

Gordon Michell
for the
Royal Fine Art
Commission

with a Foreword by
Norman St John-Stevas

The Architectural Press: London

First published in 1986 by the Architectural Press Ltd,
9 Queen Anne's Gate, London SW1H 9BY

© The Royal Fine Art Commission, 1986

British Library Cataloguing in Publication Data

Michell, Gordon
 Design in the High Street.
 1. Stores—Retail—Great Britain—
 Design and construction
 I. Title
 725".21"0941   NA6220

ISBN 0-85139-159-1

All photographs were taken by the author

All rights reserved. No part of this publication may be reproduced, stored in a retrieval system, or transmitted, in any form or by any means, electronic, mechanical, photocopying, recording or otherwise, without the prior permission of the publishers. Such permission, if granted, is subject to a fee depending on the nature of the use

Designed by Peter Ward

Typeset by Gatehouse Wood Ltd., Sevenoaks

Printed and bound in Great Britain by
Biddles Ltd, Guildford and King's Lynn

CONTENTS

FOREWORD
7

ACKNOWLEDGEMENTS
11

INTRODUCTION
DESIGN IN THE HIGH STREET
13

CHAPTER ONE
THE PROBLEM
17

CHAPTER TWO
THE SHOPFRONT IN CONTEXT
41

CHAPTER THREE
THE SEARCH FOR SOLUTIONS
71

CHAPTER FOUR
ENVIRONMENTAL IMPROVEMENT
93

CHAPTER FIVE
THE FUTURE OF THE HIGH STREET
115

This study was made possible by
**The Boots Company plc
Capital and Counties plc
Debenhams plc
Grosvenor Developments
Habitat/Mothercare
Marks & Spencer plc
J Sainsbury plc
W H Smith and Son Ltd
F W Woolworth plc**

FOREWORD

**The Rt Hon Norman St John-Stevas PC, MP
Chairman, Royal Fine Art Commission**

Architecture – like God – is not something 'out there' or even 'up there', but rather an art which has an immediate, profound and lasting effect on people's lives. The fact that its influence is not always consciously realised or reflected upon in no way alters the truth of the matter. Shopping as an activity does not at first sight seem to have a great deal to do with architecture, but if it is to be the pleasant and enjoyable experience which it should be, the architectural setting in which it takes place becomes of major importance. Well designed buildings and sensitive planning are crucial to making shopping fun; equally, overcrowded narrow streets, harassment by motor cars, ugly car parks with inadequate entrances, beetling roofs and hairpin bends can turn it into a penance and hazard.

Over the past twenty years shopping has undergone a revolution with the multiplying of shopping centres and the expansion of high streets. Too often, alas, visual and aesthetic values have not kept pace. To single out only one of many examples quoted in the report, cluttered fascias with poor lighting, oversize lettering and garish colours have been allowed to disfigure old and new buildings alike. High Streets have been allowed to become the setting for a visual shouting match which destroys not only beauty but defeats the very commercial ends which it is intended to promote.

The Royal Fine Art Commission, before whom many shopping development schemes come, has long been concerned about the problems they raise. We found that our anxieties were shared by a number of national retailers and developers and, on the initiative of the Commission, nine of them came together to sponsor this report. We are grateful for their financial generosity and their enlightened approach to the project. Gordon Michell of Michell & Partners undertook the investigation and he has produced a stimulating and brilliant report which the commission has studied and discussed carefully and which commands its enthusiastic support.

FOREWORD

An outstanding feature of the report is its principled and constructive tone. The horrors have been singled out but only so that they can be replaced by designs and objects of beauty. Our hope is that the report and its ideas will be disseminated and exercise widespread influence, not to impose a suffocating uniformity but so as to inspire good design adapted to individual circumstances and needs.

What we are looking for is a major upgrading in the quality of individual shops and radical improvement in the relevant environment. We believe that the local authorities have a leading part to play in bringing this about and Parliament has conferred upon them adequate powers to do so. Unfortunately, the powers have not always been exercised with vision and insight. It would be unfair to lay all the blame on local authorities: responsibility must be shared by traders and developers. One of the most important of Mr Michell's recommendations is for local authorities, merchants and amenity organisations to cooperate together to set up street associations, imbued with a determination to banish what he calls 'the widespread indifference to the environment we find all around us' and to substitute a real and continuing concern.

The report goes specifically into the need for creating pleasant open spaces, using planting imaginatively to enhance streets and centres, improving window displays and giving everything variety and sparkle. But it is not the aim of the report to dogmatise about details. What the Commission hopes is that the report will be carefully read by all those in a position to influence development and that local authorities will take the initiative and come forward with proposals for setting up pilot projects in different parts of the country. A report like this is the beginning not the end of the argument. I have great pleasure in recommending it on behalf of the Royal Fine Art Commission.

ACKNOWLEDGEMENTS

I would like to thank, in particular, the Steering Group made up by Mr Ian Chidlaw (Marks & Spencer), Mr Jack Gant (Boots), Mr Ian Northen (Capital and Counties), Mr David Swann (Debenhams), Mr George Reid (Woolworth), The Countess of Airlie and Miss Elizabeth Chesterton (Royal Fine Art Commission) and, finally, Mr Sherban Cantacuzino and Mr Richard Coleman, Secretary and Deputy Secretary, respectively, of the Royal Fine Art Commission. I am also grateful to Mr Julian Smith (W H Smith & Son) for his considerable help, interest and encouragement.

Officers in the planning departments of local authorities, individual retailers, shopping centre managers, designers, officers of Street Associations, members and officers of the British Retailers Association – all of these, and more, have given their time, which they could ill afford, in answering my questions, discussing the problems and helping me to appreciate the scale of the problem. I am indebted to them all for their assistance.

Gordon Michell

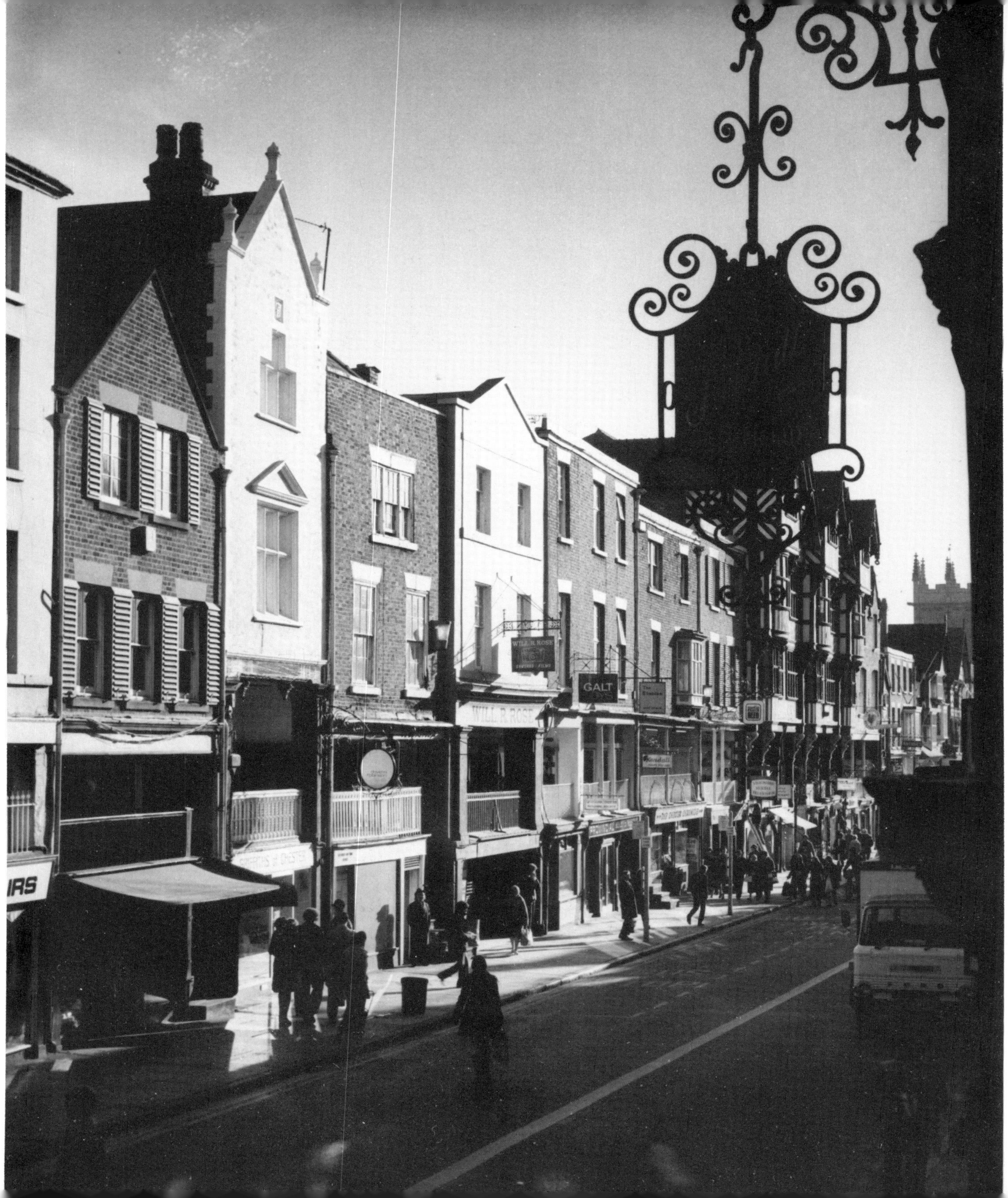
WILL R. ROSE
GALT
IRS

INTRODUCTION
DESIGN IN THE HIGH STREET

Bridge Street, Chester.

It is by their centres that most towns are remembered. These centres of commerce in village, town or city tell the story of urban development and indicate the varying prosperity of the place through the ages. Towns were seldom designed and built at one time: they have grown, been extended, altered or torn down according to the needs, the fashion and the prosperity of the day. A total lack of uniformity is almost always the rule; the resultant mixture of styles and materials contributes to the richness of our town centres. There are exceptions, of course, as can be seen in Newcastle-upon-Tyne or in Bath. The town centre is the treasury of its current prosperity and also of its heritage. The twentieth century has been less than sympathetic to the commercial streets

DESIGN IN THE HIGH STREET

Newark

Calne

and centres of our towns and has seen the erosion of character which is so important to a town. In many cases, all has been swept away in the interests of carrying out what has been misleadingly called 'urban renewal'. Today, nearly twenty years after the peak period of this renewal process, there is general agreement that the quality of urban environment has suffered a loss and that the gains of redevelopment

DESIGN IN THE HIGH STREET

The twentieth century has certainly not been kind to Newark or Calne, where a handful of stone-faced buildings is all that remains of a high street sacrificed in the interests of a 'relief road'. This gave an opportunity for new development, obviously determined to express the standards of the 1960s in its own way, regardless of the character of Calne's high street.

Amongst its many contributions to our heritage, the nineteenth century provided us with many examples of glazed shopping arcades forming spaces, cathedral nave-like in size and character, similar to this one in Leeds (right).

have not made up for the losses. In addition, where change has taken place through alteration rather than complete redevelopment, the general standard and quality of the new work has rarely contributed beneficially to the street scene.

The Study

This last sentence is an important one; it is the aim of this study to suggest ways and means by which the standards of design in the high street might be improved. The high street will be looked at in a broad sweep which will include shopping arcades, the market place and the main commercial streets of our towns, irrespective of their being high streets or not. Any attempt to base the study on the conservation of historic buildings or areas has been avoided; the study is as much concerned with the quality of the high street in towns with little architectural or historic interest. Is the quality as low as one has been led to believe?

CHAPTER ONE
THE PROBLEM

The Street

No two streets are alike. Each is made up of a variety of buildings which have been added in an apparently haphazard way over the years. Some indeed have been there for longer than we would imagine, having had later façades added to their earlier structures, chiefly during the Georgian period. The important factor in the high street is the unit; there are dozens of small units placed next door to one another and each unit, by virtue of its design, is recognisable. The design of the street generally has a strong vertical emphasis, strengthened by the fact that there is seldom any consistency of horizontal line at roof level. Exceptions to this occur where a whole block has been developed at one time and has been carefully designed to provide for shops at ground level with residential dwellings

P
KING BROS.
FROSTS
WINES AND BEERS
FINE FOODS
DELICATESSEN
VIDEO
FILMS & GAMES
Dixons
OPEN
7 DAYS
NBP 786Y

THE PROBLEM

on the upper floors. Such developments date from the late nineteenth century and continue through Edwardian days to the 1930s and 1950s. They are never on the scale of a whole street but can often occupy a substantial block on one side of the street.

Fascias

Any study of our existing high streets could not fail to underline the lack of relationship between most shopfronts and the buildings into which they are set. This can result in the whole upper part of the façade above the shop being covered by a new material which is completely foreign to the rest of the street. This 'curtain' covers existing windows, architectural decoration and features, and destroys the architectural unity of the street. In other cases unnecessarily deep fascias will obliterate architectural detail at first floor level and will cover over the sills of the first floor windows and, in the most extreme cases, half of the first floor window itself. It will also obliterate attractive architectural

Putney high street. Gasoline gulley or shopping street?

DESIGN IN THE HIGH STREET

THE PROBLEM

Any trace of the earlier windows, with their moulded reveals and pleasant decoration above, has been obliterated by the applied curtain, while unnecessarily deep fascias cut into the size of first floor windows.

DESIGN IN THE HIGH STREET

This shopfront in Gloucester, on one of the main approaches to the cathedral, has little regard for the building into which it has been inserted. Was the work undertaken with planning consent from the City Council? If not, why did the Council not proceed with an Enforcement Notice to have it removed?

THE PROBLEM

In Reading (above) and in the King's Road, London, the aggressive fascias deliberately ignore both the design and structure of the buildings of which the shop is part.

Chesterfield.

decoration on the line of the party wall. If the shop occupies the ground floor space of two or more adjoining buildings, a continuous fascia will proclaim the extent of the shopkeeper's empire by running across the façades of all the buildings concerned, again regardless of architectural detail and decoration which is hidden from sight. It is more by accident than design that a fascia will relate with its

THE PROBLEM

neighbour's; neither its material nor colour will bear any relationship to that of either side or to the materials of which the upper part of the building is constructed. This apparent desire to be different is advanced further by the confusion between the different types of lettering, and its lighting, on the fascia. More than anything else in the design of shop fronts, it is the appearance of the fascia which jars, and so degrades, the overall appearance of the street.

The Shop Window

Below the fascia line an awning, which protects the contents of the shop window from the sun, is sometimes installed; increasingly popular are ranges of small plastic hoods, often rounded or arched in shape and usually brightly coloured. The shop window with the main entrance doors are usually clear glass but often the point of this is marred by inside window stickers advertising a special offer to be found within. The glazing is frequently taken right down to pavement level where

DESIGN IN THE HIGH STREET

There is no escape from the need to maintain the appearance of the shopfront, but careful design and choice of suitable materials will do a great deal to make the task easier.

it is splashed by rain and dogs, but, alternatively, it can be terminated by a stallboard some 300–600 mm above pavement level. The shop window is more often than not one single sheet of glazing in a frame of hardwood or metal, often bright aluminium. When these frames go down to pavement level they collect dirt and, with the moisture which runs down the face of the window, it is not long before

THE PROBLEM

Window stickers on this scale are not uncommon but never fail to degrade the appearance of the shopping street and, indeed, of the shop itself.

corrosion sets in. Finally, the gap between the bottom of the trader's shop front and the back edge of the local authority's public footpath is in-filled with some material which can be pushed in to the varying space: tarmac, concrete, granolithic and terrazzo are commonly used. This dull but necessary catalogue is intended to be a fair description of the average shop front in our main shopping

The mixture of different paving materials brought together through a complete absence of design or thought is very detrimental to the appearance of a shopfront.

The lack of careful design is evident in a shopfront which is uncared for, but is nevertheless providing the display setting for exciting and colourful clothes.

streets. All too often, shopfronts are applied regardless of the design of the building above and they frequently obliterate details and dates which might be worth retaining, and including, as part of that shopfront.

THE PROBLEM

Upper Floors

Above this, windows reveal floors of vacant, under-used or mis-used space. Since there is no access to it other than through the shop, this is often used for storage which can be seen from street level. Consequently, the windows are not kept clean and the appearance of the upper part of the building clearly expresses the under-use as well as the general lack of care and attention the space receives. Where separate access to the upper floors does exist, the space is frequently not used advantageously either because the ground floor user would, for security reasons, prefer it to remain unoccupied or because of the expense of putting in the services and facilities to bring it up to the standard required for letting. In addition, owners frequently cite the poor external environment as being a deterrent to residential use.

The Environmental Quality of the High Street

The problem of the shopping street, however, is much greater than that of the individual shop unit or the sum of all the shops which make up the street. Environmental quality is so lacking in most streets as to make shopping unpleasant; in these circumstances can one blame retailers for their apparent lack of concern about the appearance of their shop, even though they are obviously interested in the appearance of the goods they offer for sale within?

The root cause of this poor environment is the conflict between feet and wheels. The very existence of the shops lining both sides of the street naturally attracts people to go there. This is the retailer's reason for being in the high street; his main aim is to attract people to his shop and to persuade them to buy more than they ever intended to buy. Shopping should be fun and a pleasurable experience. What fun is there in waiting in cold or wet weather to cross from shops on one side of a busy high street to

THE PROBLEM

those on the other until there is a gap in the seemingly endless stream of traffic? What fun is there in being marshalled across the street in two stages, the journey being controlled by pedestrian light signals and the 'reserve' in the middle of the road being surrounded by a high fence? The situation becomes increasingly unpleasant according to the amount of shopping being carried; if a pram is being

Not only is it difficult to cross the average high street, but the paraphernalia created to prevent people from crossing, to tell them when they can and to marshal them in the middle, is invariably ugly and erected regardless of its visual effect.

DESIGN IN THE HIGH STREET

pushed with one hand, with a toddler in the other, the situation becomes intolerable. That there is 'no fun' in this is true but it is only half the truth. The other half is that this continual stream of traffic of all types – motor-cycles, private cars, taxis, buses and heavy and ever larger lorries – constitutes a real danger to life and limb. One cannot window shop without the threat of an accident. One cannot walk down the street and talk because one's speech is drowned by the noise of traffic; communication with one's children is made at the top of the voice. This is all unpleasant for the shopper but the retailer himself objects, with good reason, to the dirt and the atmospheric pollution caused by the vehicles. In wet weather, dirty water is splashed across the pavement on to shopfronts. When this is mixed with salt and grit it becomes impossibly difficult and expensive to keep the shopfront cleaned and maintained. In dry weather the traffic stirs up and produces dust which makes daily maintenance of any display in the shop essential if it is to continue to look attractive.

THE PROBLEM

The Infrastructure of the High Street

In this conflict between feet and wheels, the irony of the situation lies in the fact that the wheels are necessary to get shoppers to and from the shops. The infrastructure on which a successful shopping street depends is often inadequate so that much frustration is caused by the difficulty in finding anywhere to park within reasonable walking distance of the shops. Those using public transport frequently point to the system's failure to service the main shopping street adequately. And for their part the complaint most often voiced by traders concerns the lack of adequate arrangements for deliveries to their shops and the restriction of deliveries very often to certain hours of the day. Servicing shops from the front through the main sales area is unpopular and will become increasingly so as more emphasis is placed on the quality of display and the move away from straight rows of manned counters towards a free disposition of merchandise on racks.

Environmental Care and Maintenance

A third matter constantly arises in any discussion about the present shortcomings of most high streets: this is the lack of environmental care shown by most local authorities. It can be seen in two different ways.

First, the actual ground surfaces are often ugly and uneven; when dug up by statutory undertakers in order to gain access to services below, they are all too often replaced with macadam and seldom finished with material which matches the existing surrounds. There are even examples in which a trench has been cut through red macadam and then been filled and surfaced with black.

Second, the amount of litter, coupled with the poor quality of clearance, more often than not tops the list. It accompanies general complaints about the varying standards of street cleaning, the emptying of litter bins and the length of time which apparently has to elapse before essential maintenance and repair work is carried out. There are, in addition, various amenities which people quite naturally

THE PROBLEM

Essential repair and maintenance work should be carried out without delay.

expect to find but very often do not. They would appreciate having more planting, allied, perhaps, to more seats; they want more public telephones and to find them in working order; the same applies to toilets.

Pleasure in Shopping

The importance of 'pleasure' in shopping must be further emphasised. Shopping can be a really tiresome bore, made all the worse by the noise, danger, dirt and fumes of traffic. Equally, it can be made into a pleasurable and exciting experience in which case the shopper is likely to spend more time shopping and will spend much more money as well. A sense of relaxation plays an important part in creating the pleasure – the opportunity to linger and to wander is very important. And, of course, the high street, like the market place, needs variety, must give the shopper choice and must therefore contain a keenly competitive spirit. To this must be added bustle, excitement and sparkle, for monotony is the trader's worst enemy.

Sensitive design, care and an appreciation of the street as a whole can provide pleasurable shopping, with the result that people will positively enjoy shopping there, have a sense of belonging to the place, understand its layout and, most important of all, want to return.

Milton Keynes. Architecturally interesting and exciting but otherwise lacking so much of what both shoppers and retailers require. Architects: Milton Keynes Development Corporation.

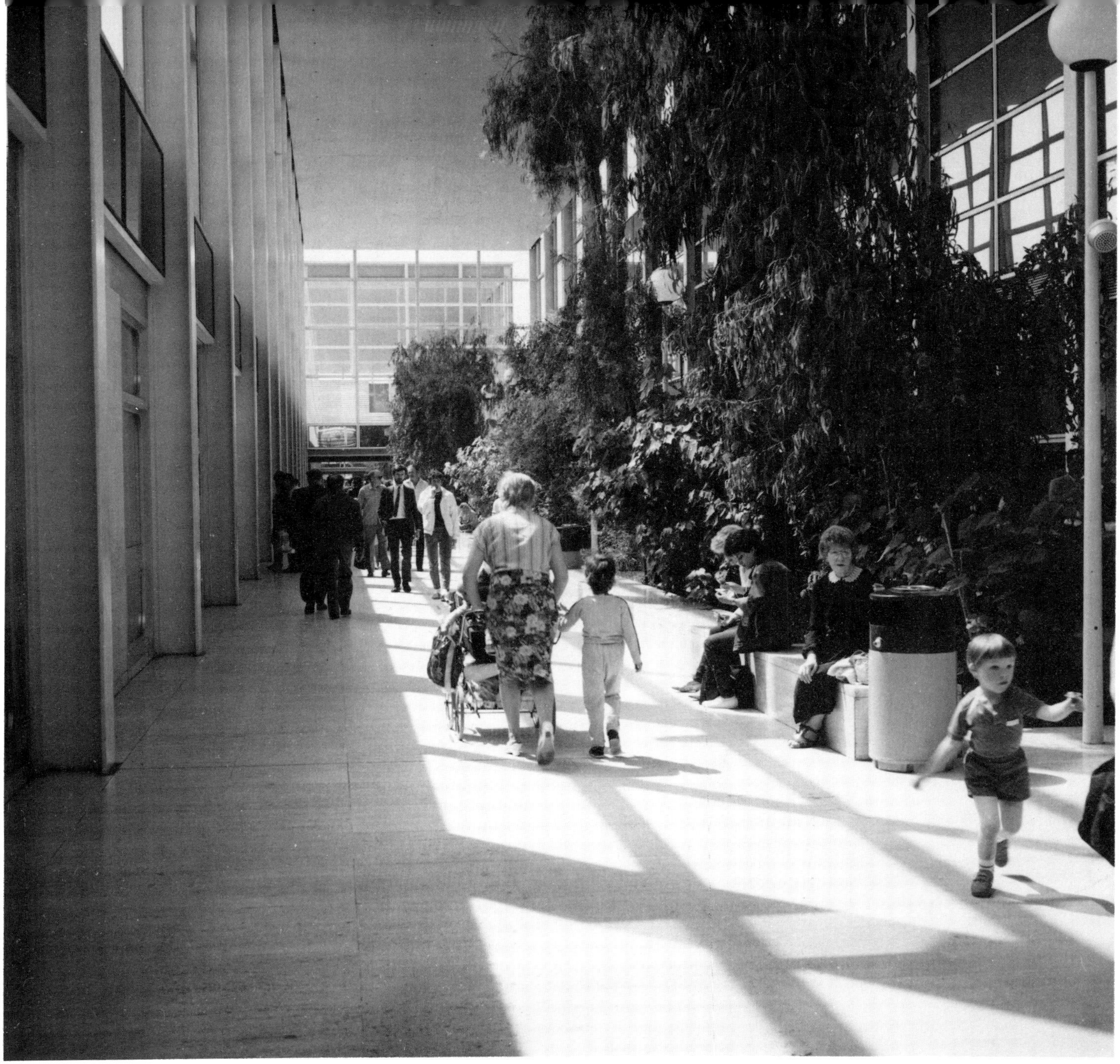

DESIGN IN THE HIGH STREET

The Total Problem

This study of existing high streets reveals an unsatisfactory picture of shopfronts unrelated to the buildings which house them and to the street generally. It also reveals an infrastructure inadequate to cope with the current needs of either shopper or trader. Finally, it reveals an alarming indifference to our environment. These three strands may be pulled together to provide the main problems to be overcome in improving design standards. It is clear that the high street must offer pleasure, accessibility and convenience; it very rarely does so, with the result that trade suffers and the larger retailers cast around for locations outside town centres in which to build anew. This can only further emphasise the importance of this study.

Time never stands still, least of all in the high street, which must speedily come to terms with the changes taking place. Perhaps now, as never before, change is sweeping through the retail industry as well as through the ranks of shoppers. More leisure time for the majority provides the

The transparency, glitter and panache of a shop in Eldon Square, Newcastle — together with the reflective mirror façade of a shop in Westgate, Wakefield, which offers a new view of the Cathedral —

THE PROBLEM

emphasises that change is sweeping through the retail industry and that this presents an opportunity to improve. Architects: Braithwaite and Jackman (Wakefield); Chapman Taylor Partners (Newcastle).

industry with the opportunity to make shopping a real leisure pursuit. There are other influences, too; the growing demand for late night shopping, the prospect of Sunday opening and the increasing number of married women at work. All of these, together with the increased spending power of the shopper, present the high street not only with a challenge but, more important, the chance to bring itself up to date.

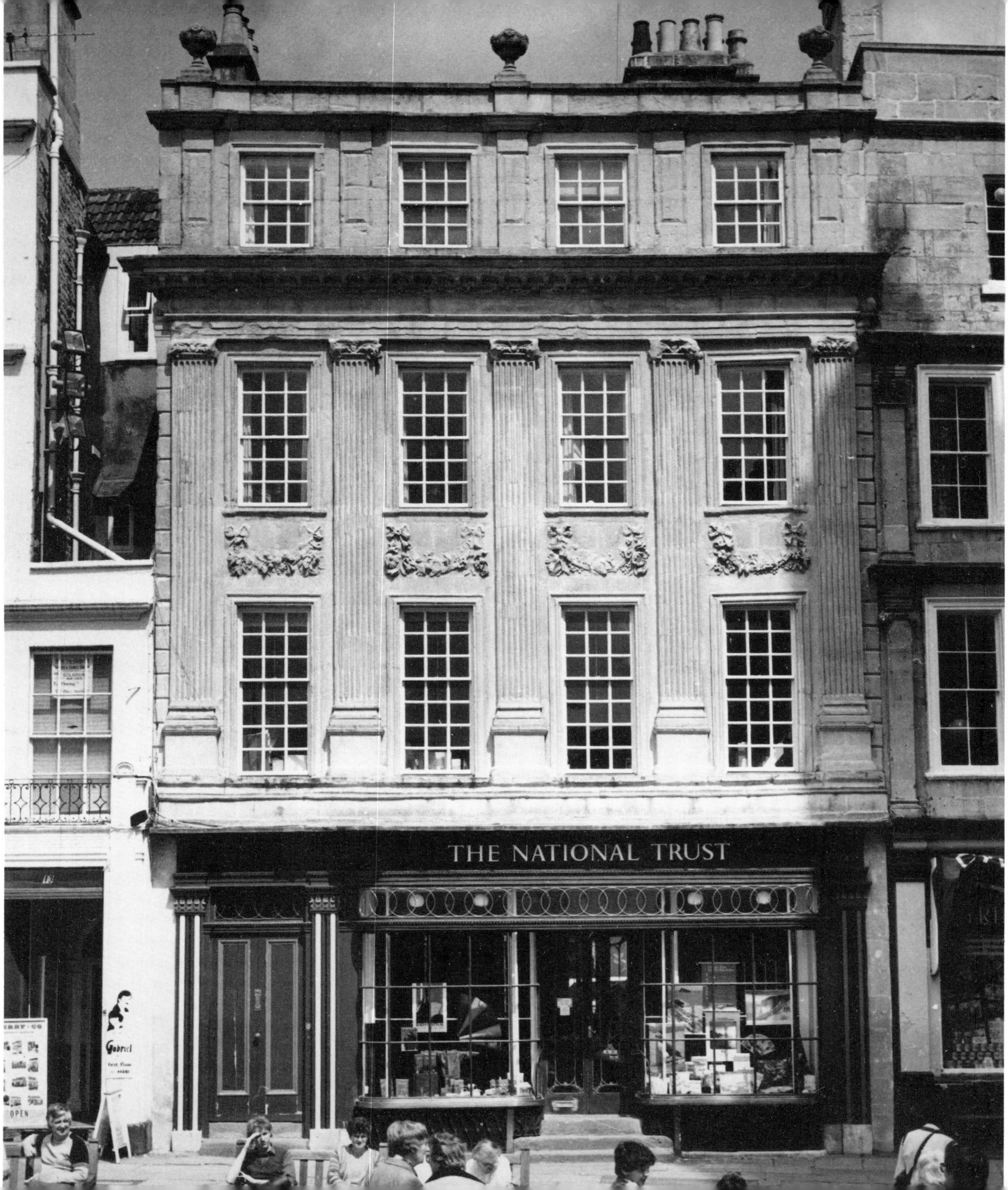
THE NATIONAL TRUST

CHAPTER TWO
THE SHOPFRONT IN CONTEXT

The Shopfront in the Street

This study has identified the poor environment of the high street as being the cause of generally low design standards. It has seen this as the essential issue to be tackled and on which to make progress. Assuming our willingness and ability to do something positive about the environment, it follows that attention should now be turned to the problem of shopfront design.

A re-statement of the problem is unnecessary. The visual chaos caused by the insertion of shopfronts, unrelated both to the buildings which house them and the street to which they should contribute, is in every high street for all to see. We can also see that shopfronts, advertisements,

Marshal Wade's house in Bath. Architects for restoration of façade: Graham and Stollar Associates.

signs and all the associated details can have an effect, often very detrimental, on the appearance of an individual building as well as on the character of the street as a whole. It is true that it can be very detrimental but it is equally true that the provision of a new shopfront gives the designer and shop owner an opportunity to add constructively and creatively to the attractiveness and liveliness of the street.

Designing for Today

How does a designer or a trader set about the task of making a really positive contribution to both the street scene and the building itself? It should be stressed that this is no conservation exercise recommending architectural pastiche. The whole business of retailing, of what is being offered for sale, its manufacture, delivery to the shop unit and its eventual use is essentially modern. There is seldom a case for designing the shopfront in anything other than a modern style; there is, however, a strong case for retaining and putting to good decorative use any

The retailer is in the high street with one objective — to sell as much as fast as he can.

elements of the original building still remaining in the framework within which the new shopfront is to be fitted.

It is also important to remember that the trader is only contemplating going there to trade successfully. He will not necessarily expect, or be prepared, to make a

DESIGN IN THE HIGH STREET

**This shopfront acknowledged the constraints of the building and accepted the need to retain the existing pilasters and cornice. The care and attention given to the detailing of the shopfront and its step, the seats at either side, as well as a handle to which to fix the dog lead, sets it far ahead of most design in the high street. However, the local authority refused it planning consent because it was considered detrimental to the appearance of the conservation area.
Designer: Dinah Casson.**

constructive contribution to the appearance of the street simply for the benefit of that street. It will be necessary, therefore, to establish that, if he does contribute positively to the street scene, his trade will flourish to a greater extent than would otherwise have been the case.

Design is, after all, simply a means of communicating a message to people, which is what all retailers must want to do. 'Good design' cannot be confined to the external face the retailer presents to the world through the appearance of his shopfront. It must be a comprehensive and coordinated approach to everything the shopper sees – the advertising, the paper wrapping or carrier bag, the shopfront and window dressing, the interior, the graphics, how the staff are dressed and so on. This will often be focused on the age or social group whom the trader is especially trying to attract to the shop. It is impossible as well as foolish, therefore, to attempt to lay down rules about the detail or design of the shopfront. This must cast doubt on the wisdom and effectiveness of some of the design guidance notes prepared by local authorities which might, by the very rigidity of their suggestions, threaten good creative design. Good creative design is essential in the high street today; it does seem possible to make a number of general points which could probably enjoy universal application.

The Shopfront and its Building

Most important, the designer of a successful shopfront will have to take account of the whole building, how it was

DESIGN IN THE HIGH STREET

The shopping environment of Leeds has benefited through the adaptation of this Church Institute to shops and offices.

originally designed and how it has changed. An objective appreciation of the building and the street should be a minimum requirement of any designer, to show that he or she really understands the building. This implies the need for someone with training to be responsible for the design of this which will not, at first, appeal to the smaller retailers who normally go direct to a firm of shopfitters. Designers

Carefully detailed, the retailers keep their identity low key, relying on the window display to persuade shoppers to enter. Architect: Hadfield Cawkwell Davidson and Partners (Sheffield).

THE SHOPFRONT IN CONTEXT

should understand the building to the extent that they become sympathetic to it without necessarily liking it. Our opinions and fashions are, like much else, constantly changing and what we have little time for today may well be popular in the years ahead. In addition to this, the designer should have in mind the character and quality of the street as a whole, since a good shopfront will not only satisfy the shopkeeper's aims but will also enhance the street scene. This is most likely to be achieved if the designer is able to control the scale of his insertion and to use materials which have an affinity with those used in the building and in the street generally. To this end, in particular, the texture and colour of the materials used are important. What needs to be emphasised is that the general tendency to try to shout more loudly than neighbours only results in everyone shouting and no one person being heard at all. Nothing but discord results from a shopfront with garish, deep fascia, projecting illuminated name box, main windows festooned with stickers advertising a range of new offers, the whole set back within, but unrelated to,

DESIGN IN THE HIGH STREET

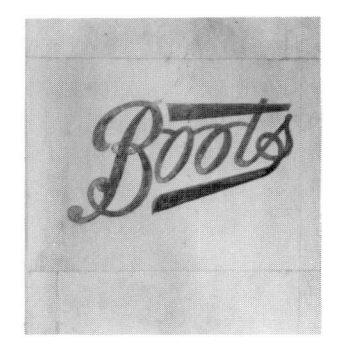

This retailer sees no need to shout in Trowbridge, while customers are given a sense of quality, permanence and security by the engraving of the retailer's name into the stonework of the reveal to the shopfront.

the structure of a carefully detailed mid-Victorian terrace of stock bricks, stuccoed window surrounds, cornice and decorated string course below the first floor windows. To make matters worse, it is doubtful whether it is even providing a satisfactory setting from which to sell the trader's goods.

The points stressed in the last paragraph may be applied equally to the design of new buildings within the high street, an issue which does occur frequently although it is not at the heart of the problem. Massive redevelopment of existing high streets has occurred, often with environmentally disastrous results and much controversy, at local as well as national level. The reason for this can usually be traced back to the developer's choice of designer; more often than not the designer's track record is enough to indicate trouble ahead. This trouble shows itself in the total lack of understanding in the new development of the setting into which it is to be inserted. The designer no doubt understands the retailer's requirements and these appear to dominate every other issue. But these requirements are not as inflexible as they might appear, as can be seen in the new shopping development of Chesterfield and Stratford-upon-Avon; in both towns, major retailers have designed their shops to fit into the town. In Stratford, Marks & Spencer occupy the old coaching inn, have foregone their normal plate glass windows and fascia and yet report that it is one of their most successful stores. Here, too, even Woolworth have dispensed with their usual fascia and lettering and rely on applied gilded and raised letters on an unpainted timber fascia. Such departures from the standard are very welcome.

The Need for Quality

All too frequently one is depressed by the poor quality of workmanship and detailing in shopfronts. This general lack of quality, and of caring about detail and appearance can be traced to the low standard of environmental design in the high street. If cracked and broken paving slabs are not renewed or are simply replaced by black tarmac; if planting beds

THE SHOPFRONT IN CONTEXT

The lack of environmental awareness in the high street is often appalling. Local authorities must make the first move in raising standards of design generally by attending to paving and by reducing the range of barriers, posts and objects which adorn the footpaths.

are not watered, tended, dug over and kept clear of litter; if the broken bench seat is not repaired and if the whole scene is dominated by ugly lighting columns and lanterns and a duplication of signs telling you what you may or may not do; if there is evidence of indifference of this nature all around, it is small wonder that retailers too feel that there is no need to spend time and money on maintaining quality and bothering about details.

Pedestrian Scale of Movement

When our high streets were major traffic arteries with vehicles passing through at anything between twenty and thirty miles per hour, it was understandable that retailers should be attracted to shop units into which huge sheets of plate glass could be inserted. Quite reasonably, retailers took the view that their potential customers had but the briefest moment in

This new house style seen in Goole has not appreciated the needs of the high street nor the building in which the store is located.

which to take in what was on offer before they had passed by and a competitor's window had come into view. Therefore the bigger the size of the shop window, both in length and height, the better it would be for trade. Nor did this craving for bigness stop at the size of the window; it applied equally well to the size of lettering on the fascia as well as to the depth and colour of the fascia. 'You've got to grab their attention at once because there won't be another opportunity once they have sped by' must have been in the minds of anyone trading in the days when the high street was little more than a traffic artery. Pedestrianisation, wherever it has been achieved, has ended all that. It has meant that every shop window is now subjected to scrutiny by people passing at two or three, not twenty or thirty, miles per hour. At a stroke the scale of the whole of the street has been reduced: the high street *should* take on a much more human scale. Big shop windows can be boring. There is now a need to break the scale down, to subdivide the window and to have more variety in the display.

A new shopfront in the Market Place at Chesterfield. On the opposite side of the Market Place an extension of Marks & Spencer illustrates how the large areas of plate glass of earlier shopfronts are giving way to smaller shop windows and larger entrances. In this case, the window dressing is minimal so that the view of the interior of the store becomes the window dressing.

THE SHOPFRONT IN CONTEXT

The Detail of the Shopfront

This reduction of scale has one further implication in that everything is looked at much more closely, for longer and in more detail. This, of course, should be welcomed by the trader; if the goods are worth buying they ought to be worthy of close scrutiny. But the ability to look at goods in shop windows also applies to the detail of the shopfront, its workmanship and its finish. Good quality merchandise in the window, attractively labelled and in eye-catching boxes with appealing colours, all deserve to be set off in an appropriate display within the framework of a well-designed shopfront.

This framework will not be achieved by using lengths of highly polished white metal with badly cut and fitting corners, or of wood frames with badly mitred corners. Yet this is all too common.

The Interior as Window Display

However, what the shop looks like in the high street will also be heavily dependent

DESIGN IN THE HIGH STREET

People moving at walking pace have time to appreciate the quality of a shopfront as well as what is displayed within. These two examples from Bath indicate that a carefully designed shopfront can provide the right setting for a smart and attractive display.

upon the trader's attitude to window display and to window dressing generally. The point has been made that pedestrianisation could and probably should result in a reduction of glass size. In removing a major source of dirt, noise and fumes, it is possible to foresee a much greater part of the shopfront being open to the pavement. An impressive feature of many shops within an enclosed centre is

THE SHOPFRONT IN CONTEXT

that they very often do not have a shopfront at all. Most of the front, if not all of it, is open to the mall and the view one gets of the interior of the shop – with its merchandise on racks and shelves – becomes the display. Thus, there is an open invitation to enter and to browse. With an increase in pedestrianisation and with the possibility of having part, if not all, of the street covered, it is clear that more and more shops in the high street will open up their interiors to the street, and display their goods in the same way. Some already do this without having the benefit of covering outside the shop; they rely on a curtain of warm air through which shoppers pass on entering the shop. It seems certain that the old style of window dressing and display will be swept away in favour of a highly selective display in the foreground, allowing a longer view into the centre of the shop or store.

It has been stressed that shopping in the high street should be fun, colourful, interesting and stimulating. The interiors of shops as well as their façades can be all of these without causing disruption to the architecture of the high street.

This repetitious display no longer belongs to the retail world of today.

It is impossible to display in the shop window everything on sale within, so it is necessary to select a small range, display them with imagination and invite the shopper to look past them into the interior.

In the pedestrianised areas of Leeds retailers are encouraged to display their goods in specially designed cases set within the paved area.

THE SHOPFRONT IN CONTEXT

This shopfront in the King's Road, London, claims attention through being in keeping with the structure of the building; it then proceeds to have an arresting display which does not fill the windows but allows a view to the interior.

The Need to Modify Standard Designs

Quite clearly, there cannot exist a standard shopfront which can be inserted in all of a retailer's outlets up and down the country. The principles of the design of the shopfront can be established but each case will need to be looked at by a sympathetic designer to see what

DESIGN IN THE HIGH STREET

This retailer has established over 200 high street outlets in the last three years. The design of the front is invariably adjusted to suit the local conditions, as can be seen in Durham (right) and Taunton (above). Concept designers: Conran Design Group.

modifications are required to insert it satisfactorily into that particular location. The same applies to other aspects of the corporate image – the fascia, lettering and use of a company logo. All must be designed so that, no matter what modifications are made, they are still instantly recognised by the shopper. Having established the basic shopfront

The loss of the matching decorating bracket at the left hand end of this fascia in Leeds is a setback. On the other hand the shopfront in Nottingham was entirely covered by a later addition which, when stripped away, revealed this fine design. Also in Nottingham is another earlier shopfront with much attractive detail, which only came to view once a later front had been taken away.

It should have been a condition of the planning consent that the continuous stone balustrade on either side of this new shopfront was reinstated.

design, what advice can usefully be offered to the trader or the designer? First, modify the standard design so that existing features of the shopfront within the original building can be retained. The retention can only enrich the new shopfront, make it more interesting and help to integrate it within the building as a whole. There are examples of shopfronts whose side piers and fascias have covered up the original features; the removal of these additions has revealed the original features and these, as well as making the shopfront more intersting, restore the integrity of the building.

Choice of Appropriate Materials

Secondly, most buildings into which a new shopfront is to be inserted are built of matt and non-reflective materials.
Polished granite and marble are an exception to this but they are less reflective than the highly reflective materials now on the market which can look out of character, particularly if used as a long and deep fascia. Much more important, however, is that if this material is to be used the design and number of joints in the panel should be very carefully monitored. Also important is the way in which it is fixed and made even in plane; being so reflective the unevenness of the fascia cannot be concealed and thus contributes to a general lack of quality in the shopfront.

DESIGN IN THE HIGH STREET

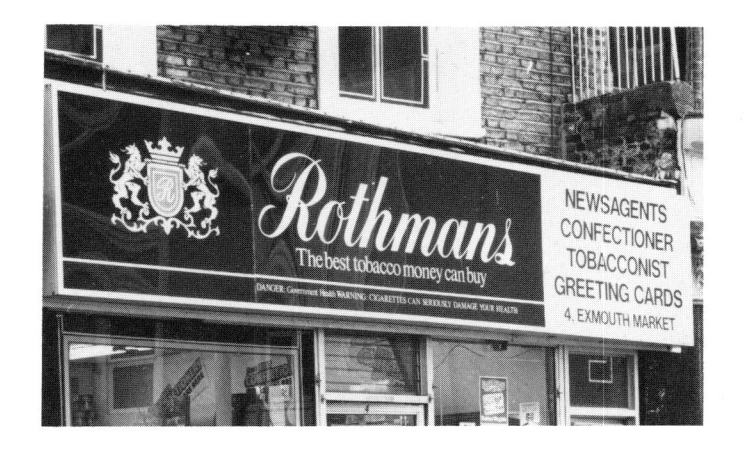

The fascia (left) incorrectly gives the impression that the shop is in the ownership of a firm of cigarette manufacturers who even have to reproduce the standard health warning on the fascia. An example (right) of an illuminated box fascia creating a new raised plane on the shopfront.

Lettering

Fascia lettering is both important and significant. Advice has already been offered against the use of boxed and illuminated fascias which simply represent another element in the façade by the creation of a raised plane. It is better either to signwrite the name on the fascia or glass of the window or to apply individual letters. A fascia which bears other advertising material in addition to the name of the shop is never satisfactory. It is also open to the trader to dispense with the name on a fascia, relying on the use of a logo to identify the shop.

THE SHOPFRONT IN CONTEXT

It is not so much the lettering as the colour combination which jars in the street scene. The lack of relationship between the letter which forms the logo and the main lettering is emphasised by the fact that the logo oversails its brightly coloured base at the top and bottom.

This fascia in Shaftesbury conveys the message just as effectively as that on page 52 and is much more acceptable in the street.

DESIGN IN THE HIGH STREET

These four examples all display a quality of design and finish which makes them stand out in the street.

Projecting Signs

The use of a well designed projecting sign can enhance both the shopfront and the street. Box sign manufacturers achieve dull mediocrity in the rectangular signs projected from the face of the building, usually at the party wall. Shops used to make such good use of hand carved and painted signs which could make both the daytime and, with today's spotlights, the evening view of the street so much more

THE SHOPFRONT IN CONTEXT

A well designed projecting sign can enliven the street scene by creating interest and amusement. Avoid the use of rectangular illuminated box signs; lighting by a single spotlight is much more effective.

interesting. Why should the brewers be almost the sole defenders of the hanging sign? Equally, why are pawnbrokers, and more rarely barbers, the last traders in the street to advertise their existence by hanging or projecting features?

Stall Risers

It used to be common practice to design a shopfront with a stall riser. This provided protection from feet and dogs, but was also proof against rain splashing up from a

DESIGN IN THE HIGH STREET

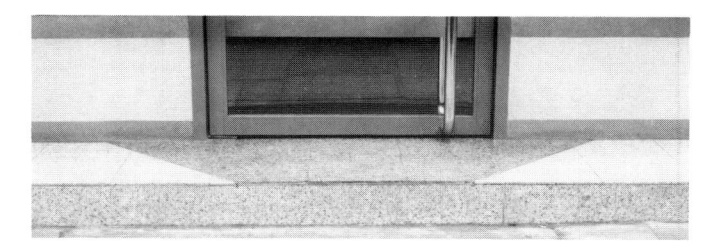

If possible, the stall riser should continue the fabric of the building so that the shopfront is simply an insertion.

dirty pavement. Practical though they still are, even today, stall risers have almost gone, although both the rain and the dogs remain.

The Role of the Local Authority

If there is general agreement about the desirability of raising design standards, the local authority planning department is best placed both to encourage and insist on these higher standards being met. The negative aspects of development control and the planning system in general are too frequently stressed; there is an important creative aspect to both which is underrated and, indeed, often frustrated by over-cautious or non-visual decision makers. Creative design in the high street should be encouraged. Mock Georgian bow windows are not what this study is advocating. The local authority could help this creative work to be successful by outlining their view as to the character of the local high street which they consider worthy of retention and enhancement. They have to be the watchdogs to ensure that an application for a new shopfront, or even for an alteration, is going to improve the high street. Once consent is given, they have to ensure that quality and excellence are being maintained in the new work. 'Quality' here does not apply simply to the finished article; it applies equally to the design. Through collective indifference, quality is all too often absent. There is evidence, in the Midlands for example, of an enlightened new approach by local authorities in cities such as Birmingham, Leicester and Nottingham. Here, generous grants have been made available to help owners of shop premises to improve not only the shopfront but the entire elevation and roof above. In this way whole terraces of formerly run-down shops have been improved, the areas have become more attractive to shopkeepers and shoppers, and the economic viability of the shops have been improved. Owners have been encouraged to carry out internal improvements and the condition of the existing stock of buildings has obviously improved. This is an enlightened approach by local government and needs more universal support by central government. It is also very much in line with the work of developers of the early shopping centres built in the 1960s where a great deal of much needed radical refurbishment is now in hand.

CHAPTER THREE

THE SEARCH FOR SOLUTIONS

Introduction

Previous chapters have dealt with the outstanding problems of the shopping street and the shops within the street. Such an approach is bound to provide an uninspiring picture but now is the moment to consider positively the real opportunities which exist.

It must be possible to produce examples to show what can be achieved if one begins completely anew. Would a quick look at newly built enclosed shopping centres illustrate higher design standards as well as an improved environment for shopping?

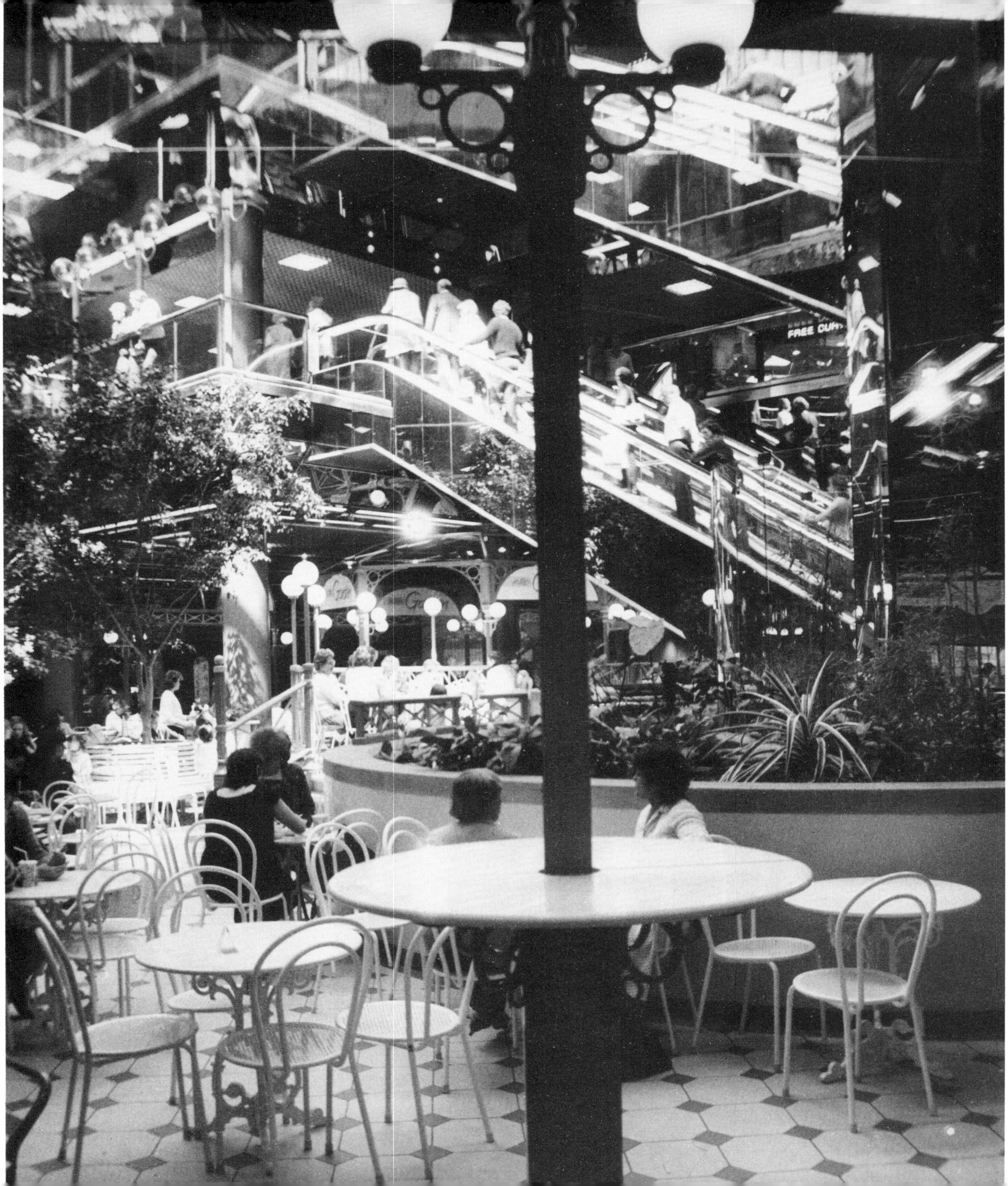
FREE

THE SEARCH FOR SOLUTIONS

Freedom from Vehicular Traffic

The immediate and outstanding impression gained from visits to such centres is concerned with the benefits derived from clearing out the traffic. Whatever shortcomings the centre might have, this freedom from the conflict between pedestrians and vehicles makes a tremendous impact, particularly after one has been in a typical shopping street with all the traffic problems already noted. It is most obviously appreciated by mothers with children but, judging by the numbers of people using these centres, especially on a Saturday, it is an asset to people of all ages.

Centralised Control

A second impression is the benefit derived from the centralised control of single ownership. This makes itself felt especially in the standard of maintenance of paved areas and the cleanliness throughout. Most centres require tenants to participate in a tenants' association, which gives them

The Ridings Centre, Wakefield. Architects: Chapman Taylor Partners.

a sense of involvement in the running of the centre.

These are two important advantages and both influence the environmental quality of the centre. Many of the early centres, designed and built in the mid-1960s, however, have two drawbacks, which sets them apart from the more recently designed centres.

Car Parking

Many shoppers get their first impressions of centres upon arrival at the car park. The quality of the architectural and environmental design of these car parks has usually been minimal, reflecting the fact that they have been built as cheaply as possible. It is frequently difficult to find the right route to follow when looking for a parking space. The floor areas are often big and the ceilings are always as low as possible. This results in spaces with inadequate natural lighting and often patchy artificial lighting. Once out of the car one has to traipse through parked cars and across vehicular routes to a ticket vending machine, a journey which must be made again in order to leave the ticket on the car before the journey back to the lift or stairs can be made. Conditions are potentially so dangerous that it is surprising there are not more accidents, particularly with children released from the confinement of the car. Even if such car parks do not have a high accident record, there is no disputing the dreadfully low environmental quality of the interior. Conditions are noticeably worse in centres where the car parking management is in different hands from that of the centre itself. These poor conditions are not confined simply to car parks and access ramps but continue to the lift halls, the state of the lifts themselves and to staircases. It comes as a relief to escape from these areas and into the shopping centre itself but, by then, the damage has been done and the first impression of the centre has been stamped on the minds of countless shoppers. It cannot be the impression the designers have wished to convey and it is one which every thinking retailer must wish to avoid, although these problems are entirely beyond their control.

**Eldon Square,
Newcastle-upon-Tyne.
Architects: Chapman
Taylor Partners.**

The standard of finishes and the high level of maintenance repair and cleanliness are striking features of most shopping centres. Those which rely entirely on artificial lighting are undeniably claustrophobic, but many are now undergoing expensive schemes of refurbishment aimed at introducing an element of natural daylight, food courts and other improvements.

**Eldon Square,
Newcastle-upon-Tyne.**

The Arndale Centre, Manchester. Architects: Hugh Wilson and Lewis Womersley.

The Value of Natural Daylight

The second drawback encountered in many of the early centres derives from the lack of natural daylight in the shopping malls. As a result they tend to be claustrophobic, dull and, particularly on a bright summer's day, appear underlit and dark. The lack of daylight also results in a lack of landscaping and planting so that nowhere is there any degree of softness; all the surfaces in sight are hard and non-sound absorbing. The designers and developers have frequently done their best to counteract the lack of light by providing fountains, big central clocks and other pieces of sculpture.

Refurbishment

With such unattractive surroundings, therefore, it is not surprising that these centres are not places in which people feel the urge to linger or to wander. There is rarely any indication of that 'pleasurable experience' already identified as one of

DESIGN IN THE HIGH STREET

Refurbishment is not confined to shopping centres. Birmingham, Leicester and Nottingham are amongst cities in which the refurbishment of terraces of shops and residential accommodation has been undertaken by local authorities as part of a scheme to stimulate trade in the local shops.

the most important ingredients of satisfying and successful shopping.

The importance of 'pleasurable experience' is now widely recognised, as can be seen from the amount of refurbishment work in hand in many of the early centres. Such refurbishment is not confined to decoration but involves bringing daylight into previously artificially lit malls, providing glazed colonnades to shop units on the perimeter, enclosing open malls, remodelling entrances and providing a new scheme of signs and graphics generally. Much of this work can be very difficult and expensive to achieve, particularly in cases such as the Victoria Centre, Nottingham, where residential accommodation is built into and above the centre which gives rise to understandable complaints about the noise of the work. In other centres, steps are being taken to provide fast food courts, but all of this work causes disruption to shoppers as well as traders, which is costly. There are, however, examples of recently built centres in which, perhaps, most of these problems have been overcome.

Milton Keynes

The new, much publicised shopping centre at Milton Keynes certainly does not suffer from the problems of the early centres. Indeed, the principle of parking cars around the perimeter on the same level as the shops appears to be ideal, although the centre has proved so popular that additional car parking has now been provided at the same level as the shops, but further away. Although people still have some distances to walk, all the parking areas have been land-

Parking amongst trees at Milton Keynes.

scaped and planted on a scale which provides instant cover and pleasure. It would not be possible to provide car parking in this way for any established centre or high street in the middle of an existing town. It is nevertheless very impressive and makes the Milton Keynes centre a nice place to be in, which is more than can be said of the car park at the Grosvenor Centre in Northampton. The environmental slum conditions persist in a park owned and run by the local authority who, on account of their wide range of responsibilities and the limitations on their budgets, are probably not the right people to provide car parking for good shopping locations.

The interior of Milton Keynes, similarly, is very different from the Grosvenor Centre and from countless other centres on account of its high, spacious and extremely well-lit shopping malls. These, too, are planted with a wide variety of trees and the environment generally is very attractive. Nevertheless, the centre has a deadness and sterility which is not good for shopping. One of the problems of its conception is that shoppers do have to walk considerable distances; the distance from a major store at one end to a smaller retail unit two-thirds of the way towards the other end and then back to the car can mount up. This distance would be easier to accept if there was more life and vigour in the malls and if shopfronts were visible down the length of the mall. But distant views are blocked by the trees in the middle of the mall, and they also obliterate views of nearby shopfronts. These are not really visible until one has almost reached the shop, since they are so deeply recessed behind the framework of the centre's architectural grid. This,

DESIGN IN THE HIGH STREET

Milton Keynes.

coupled with the fact that there are no projecting signs, makes it necessary for the shopper either to know from previous experience the layout and distribution of shops or to rely on the use of a layout plan.

THE SEARCH FOR SOLUTIONS

Milton Keynes. High, spacious and sunlit malls with generous planting and seats makes this a pleasant place. Its uniformity of colour, lack of projecting shopfronts and signs, as well as the way in which the structural grid dominates, deprive this centre of the vitality and interest which a successful shopping centre depends on.

Peterborough and Wakefield

A dead atmosphere is certainly not a charge one could levy against the Queensgate Centre, Peterborough, or the more recently completed Ridings in Wakefield. Both have about 100 retail units of varying size, both have car parking facilities for 2,000 cars (Peterborough) and 1,100 cars (Wakefield), both have retail catchment populations of about two million. Peterborough has a bus station as part of the development, with capacity for 800 buses daily. Both are located in the very heart of the city with their respective cathedrals close by and both involve the insertion of new buildings within the older fabric of the city. Both have natural daylight, the centre at Wakefield being particularly successful in getting this, and the sunshine, into the heart of the centre and at all levels. The integration of all modes of transport within the planning of the centre at Peterborough is very good and the advantages this offers to the shopper must contribute to a more relaxed attitude to shopping on the part of the shoppers. They are now offered a quiet,

The fast food court at The Ridings, Wakefield (left), at the lowest level from which escalators rise up to upper floors. On the top floor is another restaurant (top right).

One of the main shopping malls at The Ridings showing the way in which natural daylight reaches down to lower levels through openings in the floor, which also make the Centre spatially more interesting. The lack of a consistent line to which shopfronts must reach also adds to the variety of the scene.

DESIGN IN THE HIGH STREET

The Ridings.

The Ridings.

The view from the main entrance through roof glazing to the tower and spire of Wakefield Cathedral.

traffic-free, well-maintained and naturally lit environment in which to shop. Wakefield does not have the same public transport facilities as Peterborough, but it has three levels of naturally lit malls, each relating to a nearby ground level of the existing site which slopes steeply from the south northwards through the centre. Part of the development has incorporated an earlier shopping precinct of the 1970s, which has been refurbished and covered, and now appears completely unified with the new centre, the general character of which is of existing city streets, pedestrianised and covered. There is little evidence of strict architectural control of shopfronts although it is clear throughout that the developers and their architects have exerted a quiet control in order to achieve good quality of finishes, details and standards generally. The result is a centre which sparkles with natural light and sunshine, reflects interest and lack of standardisation in construction and has unbroken views. Above all, people give the impression that it is a place in which shopping has really become a pleasurable experience.

The Lessons

This review of shopping centres was undertaken in order to see what could be achieved if developers and architects were given a free hand and could create a completely new environment for shopping. What sort of environment would it be? Would there be outstanding lessons to be drawn from such a review which might be applicable to design in the high street? The answers are that there are a great many lessons in these centres from which the street must learn if it is to hold its own and prosper. These can be summarised, as follows:

1. Pedestrianisation The release of the shopper from the anxiety, noise and dirt of motor vehicles must be at the top of any list. This can only be achieved as a result of appropriate investment being made towards the provision of an adequate infrastructure for the high street. There must be reasonable access to enable traders to service their shops and there must be adequate car parking for shoppers within less than a quarter of a mile of the principal shops. Finally, there must be easy access to adequate public transport systems.

In some cases total pedestrianisation will be impossible to achieve. In such cases it will be necessary to design a scheme of traffic and environmental management to restrict entry to various types of vehicles or to various times of the day. The crucial issue is to find some way in which to reduce the conflict between feet and wheels.

2. Protection from the Weather The introduction of daylight and sun into the covered shopping mall can transform the character of the mall into an extension of the city street network. Pedestrianisation should not be regarded as an end in itself but as a first step towards the provision of more civilised conditions for shopping. The second step, therefore, would concern itself with ways in which the shopper could be protected from the elements through glazed, covered walk-ways, through glazed colonnades, or through entirely covering and glazing the existing street.

3. Tenant Mix The creation, through the

application of sophisticated tenant mix techniques, of a strong pedestrian flow. This is now well understood in the planning of shopping centres which at the outset depends on the strategic location of the big name national stores, backed up by strong representation of the smaller but high quality specialist shops.

4. Amenities The opportunity to carry out creative, attractive and varied planting and landscaping. Also to create central spaces for important amenities such as grouped telephones at one place, seats at another, a sculpture court at another, public toilets and play facilities elsewhere.

5. Management Control The existence, through single, or at least very limited, ownership patterns of a strong unified management control. The effect of this is obvious; there is a high standard of cleanliness and litter removal, maintenance of planting, design of shopfronts and, generally, freedom from window stickers; these never fail to cheapen both the individual shop, in particular, and the mall in general. The standard of cleanliness and maintenance of these centres separates the best of them from the average high street. Some centres do appear to be as ill-maintained, dirty and run-down as the high street but they are the exception. The rule is that litter is cleared up, floors are kept clean and polished, the perennial problem of paving stones being out of place or lifting at one corner is quite unheard of. How different all this is from most high streets.

If these very basic lessons are accepted, the suggestion is that the opportunities now open to us for creating a better environment in our high streets are tremendous. However, these opportunities will not be realised without vision, imagination, courage and drive. However difficult success may be to achieve, it will, in the long term, be seen as an investment for the future.

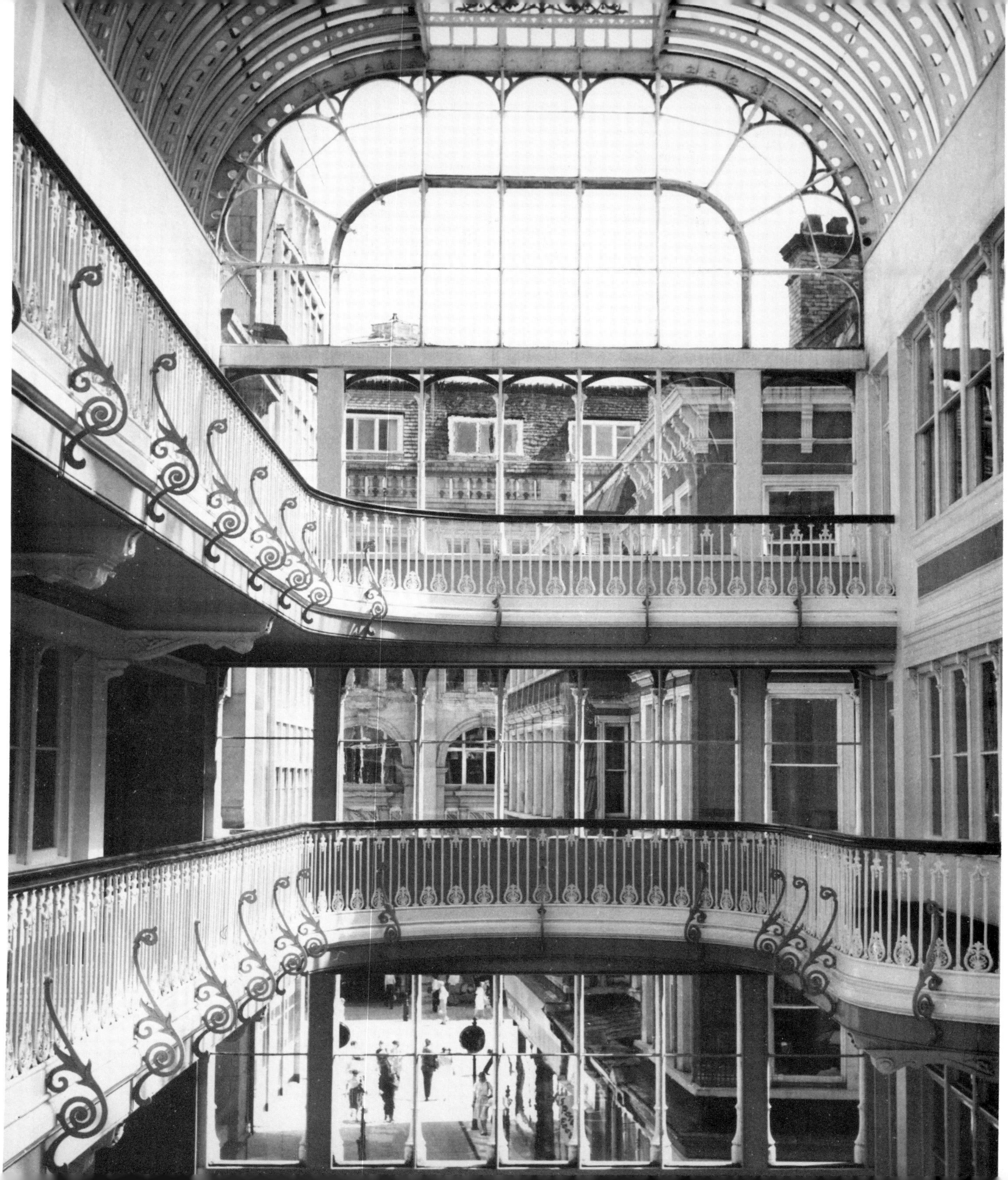

CHAPTER FOUR

ENVIRONMENTAL IMPROVEMENT

The Heritage

It is not necessary to turn to recently built shopping centres for evidence of earlier efforts to create a good environment. Much of the architectural heritage of this country is made up of examples of commercial premises where a good environment is obviously highly valued.

Glazed Colonnades

In Buxton, Llandudno and Southport, for example, cover beyond the shopfront was afforded to shoppers by the erection of glazed colonnades. Those at Lord Street, Southport, testify to the generous thinking behind such layouts – no mean covered ways but broad, glazed colonnades giving good light to shop windows and sufficiently wide for four or five people to

Barton Arcade, Manchester. Restoration architects: Turner Lansdown Holt Manchester.

DESIGN IN THE HIGH STREET

A handsome colonnade at Buxton being re-glazed.

walk down the street abreast and still be under cover. Beyond the colonnade itself runs a wide pavement, while down the middle of the street is a series of gardens laid out by the eminent landscape architect, Thomas Mawson.

ENVIRONMENTAL IMPROVEMENT

Covered Markets

In addition, we have had a tradition of covered market buildings, many of which, regrettably, have already been swept away either by redevelopment following wartime bombing or by redevelopment of whole central areas during the boom years of the 1960s. Enough remains to remind us of the loss and many are still in

The environment of Southport's Lord Street is helped by the generous width of both the colonnade and the pavement beyond.

Leeds (left) and the Penny Bazaar in Newcastle (right).

use and being refurbished, especially in the northern cities. Newcastle-upon-Tyne still has its thriving market of 1835, built by Richard Grainger to the design of John Dobson, in which Marks & Spencer established the original Penny Bazaar, which is still there today. It was considered the most spacious and magnificent in Europe. Elsewhere, in Leeds for example, marvellous iron constructions were erected to house the market and to cover and protect traders and shoppers alike. Within recent years, Londoners have witnessed the departure of the wholesale fruit and vegetable market from Covent Garden, in central London. The empty buildings have been taken over by the GLC and the whole area converted into a virtually traffic-free area, with the market buildings adapted to house a wide range of small,

The re-use of old buildings in an imaginative way, together with pedestrianisation, has transformed London's Covent Garden into a place which people enjoy. Architects: Greater London Council.

DESIGN IN THE HIGH STREET

Within the Lord Street colonnade.

specialist shops, bars and restaurants. The entire venture and the environment created for shoppers has confirmed beyond all doubt that Covent Garden offers people exactly what they want. The shops are attractive and an environment has been created for *people*. Consequently, they go in their thousands.

Glazed Arcades

Pedestrians dominate also in all those Victorian and post-Victorian arcades of which, again, all too few remain. Newcastle-upon-Tyne's Central Arcade of 1906 is a beautiful space, light and airy with a carefully designed and executed tiled floor. Another example is the recently restored Barton Arcade, Manchester, which has ground floor shops and three floors of office space above, all approached from upper level galleries which wind their way around the three tiered and beautifully top lit arcade. This example is all the better since it fits into an increasing amount of pedestrianisation which the City Council have created in the area, so that the arcade becomes part of the pedestrian network. Much the same has happened on a much bigger scale in Leeds, where several arcades have been brought into a broad scheme of city centre pedestrianisation. The now traffic-free and landscaped streets are full of people and attract major shops, while the arcades attract smaller specialist shops.

In Manchester the Barton Arcade links into a wider area network of pedestrianised streets and arcades.

In Leeds, a number of arcades have been refurbished and now form part of a central area network of pedestrianised streets and arcades.

DESIGN IN THE HIGH STREET

In this example, closure of the street to traffic has encouraged the retailer to open up the whole of one bay of his shop as an entrance.

A Pedestrian Environment

All of these examples have one common theme: people want to be in a pedestrian dominated environment. Initially fiercely fought by Chambers of Trade as being bad for business, such environments must now surely be accepted as being good for people and therefore good for trade. The list of successes is impressive: Durham,

Newcastle's Central Arcade.

ENVIRONMENTAL IMPROVEMENT

Early in the morning, before many people are about, it is possible to appreciate the care and trouble taken in paving over this main street in Durham. Designers: Durham City Council.

A more typical scene, resulting from the closure of a high street to traffic, is that of Lincoln.

York, Leeds, Wakefield, Lincoln, Liverpool, Chester and Chesterfield, Cardiff, Shrewsbury, Peterborough, Winchester, Rochester, King's Lynn, Oxford, Taunton and Chichester are but a few of the towns and cities in which successful schemes of pedestrianisation have been carried out.

Reallocating Urban Space to Provide more for Pedestrians

It would be impossible to take traffic out of every high street. Where it cannot be done, it will be necessary to investigate every opportunity to reduce it and to achieve some form of environmental improvement. Perhaps this may only take the form of widening the pavement, but most places would benefit from that alone. The sense of relaxation and pleasure gained from the generous width of pavement in Lord Street, Southport, has already been mentioned. Far too often do we find the pavement inadequate to accommodate the growing pedestrian traffic, prams, wheel chairs, and shopping baskets, as well as all the

ENVIRONMENTAL IMPROVEMENT

The problems of Oxford Street — London's major shopping street — are such as to justify more radical solutions than have so far been implemented.

DESIGN IN THE HIGH STREET

The success of Covent Garden depends, in some cases, on simple solutions — large wooden bollards which separate feet from wheels.

paraphernalia connected with traffic lights and pedestrian barriers.

In other places it may be possible to close the street by paving a section, so stopping its use as a through road. This has been done in Taunton by paving over the last fifty yards of the street, allowing the rest of the street to be used for access and parking. The traffic flow has been radically reduced and conditions for shoppers have improved, which has lead to an upgrading of the shops themselves.

In some cases the demand for more space for pedestrians appears to be almost insatiable. London's Oxford Street is a good example of an attempt being made to cut down the through-traffic, but there is still too much traffic. The draw of the shops is so great as to make it clear that more must be done to provide the street with acceptable environmental conditions to make shopping there a real pleasure.

The Economic Success of Pedestrianisation

A survey of 105 cities published in 1978 and carried out by the Organisation for Economic Co-operation and Development with Europa Nostra considered that the movement to create pedestrian zones had reached its culmination although 80 per cent of zones examined reported an intention to expand the zone. There is certainly sufficient evidence to pinpoint the economic success of pedestrianisation; a Civic Trust survey published in

DESIGN IN THE HIGH STREET

1976 indicated increases in shopping turnover following pedestrianisation as varying between 10 per cent and 15 per cent up to 40 per cent. In a survey of shoppers in London's Carnaby Street, 81 per cent considered pedestrianisation to be a good idea. There is also evidence that pedestrianisation promotes improvement of shopfronts, a change in the image of the shops as well as an increase in their floorspace. As a first step towards more ambitious proposals and to encourage the faint-hearted, why not close main shopping streets to traffic on a Saturday afternoon or on Market Day, or for three hours of a late closing evening? Guildford, for example, closes its high street to traffic every Saturday. It is an arrangement which has now operated satisfactorily for several years and, judging by the number of people who use the street, it is popular. Those who lobby against the introduction of traffic reduction are concerned with the need to provide alternative routes for the traffic, but it is interesting to note that of 146 German schemes of pedestrianisation only 59 per cent have been accompanied by new road construction. John Roberts, writing in *Pedestrian Precincts in Britain* (Transport and Environmental Studies, 1981), further reports that North American malls rarely have new roads associated with them and that much central area traffic does not need to be there. Indeed, it seems that displaced traffic frequently disappears.

These schemes demonstrate clearly the popularity of high streets after radical environmental changes have been made. Such improvements enable high streets to become really competitive with shopping centres.

Urban Landscaping

The removal, or at least reduction, of traffic is the most important part of urban landscaping but this is not an end in itself. It is also important to consider the design of the street as a whole in re-paving, planting, siting telephone kiosks, seats, litter bins, signs and notice boards. This must be undertaken as a single unified design exercise and an understanding of the composition of the street is therefore

DESIGN IN THE HIGH STREET

Glazed colonnades at Saltburn have been restored.

A new shopping scheme (left) off the high street in Thornbury has had its pedestrian ways covered to protect shoppers. Architects: Alec French Partnership.

ENVIRONMENTAL IMPROVEMENT

vital. The aim will be to provide an attractive setting for the buildings which form the high street, to offer an environment which affords both pleasure and relaxation to the shopper as well as a place in which shops can advertise their presence and their wares effectively.

Protection from the Weather

Shopping areas can be protected from the elements in a number of ways – the glazed canopies of York's Coppergate, which cover both sides of the street, offer one solution; another might be a central way covering an entire street, linked to shop entrances. A further solution, again seen in York, might be to provide a glazed covering of the entire street up to the top of the fascia, as has been done so effectively in a new development of small shops. This particular solution offers hitherto unseen views of York Minster through the glass roof. Where buildings are a reasonably uniform height, it might be worth covering the entire street and its buildings with a glazed roof, as is now

New shops at Stonegate in York have glazed roof coverings to the pedestrian ways, offering fascinating new views of the Minster. Architects: Tom Adams Design Associates.

Coppergate, York. Architects: Chapman Taylor Partners.

ENVIRONMENTAL IMPROVEMENT

being proposed for the town square in Basildon, Essex. Basildon further proposes to build a central island of shops, thus making good use of the space, enlivening the area and helping it to pay for itself. Such ideas continue the much admired tradition of Victorian glazed canopies and arcades. New fields have opened up through technical advances in the manufacture of glass fibre-reinforced materials which enable us to design covered spaces. In the hands of sensitive designers such materials need not detract from the architecture of the high street; they offer tremendous opportunities for positive enhancement. 'Enhancement' on this scale will call on the best skills of creative designers who, in many cases, will be working in a street with historical associations and buildings of architectural interest.

BASILDON
TOWN CENTRE CARNIVAL TIME
ARMY & NAVY
PATISSERIE
COFFEE BAR
Aristocrat
Sale

CHAPTER FIVE
THE FUTURE OF THE HIGH STREET

Introduction

There would have been little point in undertaking this study if it were felt that the future of the high street was limited; that it would soon become as run down, even as dead, as downtown areas in certain American cities. So far, Britain's high streets have shown no signs of following the course of their American counterparts, partly because few of our cities have been subjected to the devastating programme of urban motorway construction that America experienced immediately after the war. On the other hand, there is no doubt that the make up of our high streets will change.

The proposed covering of the Town Square in Basildon. Architects: Michael Hopkins Architects.

DESIGN IN THE HIGH STREET

The Need for Change

The retail industry itself is in a state of constant change and must continue to be so to respond to the pressures of competition as well as the needs and desires of those whom the industry serves. This need for change, alteration and modification in order to keep pace with current trends will influence designers away from the expensive and elaborate interior storefitting of the past. Once, shopfronts and shop interiors were built to last; this is no longer necessary. Adaptability, flexibility and ease of refurbishment are now the key issues. Not only will fashions change but new technology will be introduced to cut labour costs and scope for human error as well as to facilitate accountancy. Additionally, the continuing diversification of the big multiple retailers into new fields is necessarily expanding their business, increasing their demand for space and storage as well as creating additional servicing problems. Some of them are outgrowing the high street as a trading forum – the accommodation is inadequate and in many cases it will be physically impossible to provide the infrastructure they require without unacceptable destruction of the existing fabric. It is, however, important to understand the scale of likely change of this nature. The investment in our high streets is, and will continue to be, enormous. Suitable out-of-town sites will be difficult to find and planning consents might not be easily acquired. To take one example, Marks & Spencer are known to be looking outside the high street but this hardly poses a serious threat when one remembers that they currently invest in over 250 high street locations. This will not spell out the end of the high street; it will simply change its nature. It may well contract or become more concentrated; the place of some of the larger multiples will be taken by smaller specialist shops of quality. It is more likely that some of the specialist shops which have largely disappeared from the high street will come back.

A High Street Partnership

Although traders need have no fears about the demise of the high street, one senses that many have lost confidence and have been pushed onto the defensive when they should be aggressively promoting trade in the high street. The high street has a vital future which must be fought for; it will only be won by close cooperation between all concerned, including the statutory authorities. One of the recognisable lessons to emerge from a study of shopping centres was the value of a central organisation responsible for management and maintenance. A comparable organisation which represents traders, looks at important promotional issues, concerns itself with the appearance and environment of the high street – in short, cares about trade and traders in the high street – is desperately needed. How can the high street get such an organisation? What might it do?

First, and high on any list, is the need for the high street to *promote* itself. People within its catchment area must know what facilities, apart from commercial, it can offer – for instance, child-minding facilities, libraries, centres for the elderly and so on. So many high streets offer a wide range of facilities but people do not make the most of them because they do not know they exist. If the street promotes itself, it will attract more people to amenities and shops, thus giving traders the opportunity to increase their annual turnover.

Second, such an organisation could act as a pressure group in representing traders' views to the local authority over proposals for change, as well as keeping the high street clean and well maintained.

Third, it could act as a persuasive force over traders whom it felt were failing to maintain the high standards of the high street.

Fourth, it could become directly involved in some aspects of maintenance. It would be by far the best organisation to take charge of street planting; it would not involve traders giving a lot of their time to it. The presence of someone to remove litter from beds, to weed and water the soil in summer, would quickly transform the

DESIGN IN THE HIGH STREET

appearance of any urban landscape. Local involvement would also be immensely beneficial.

Fifth, its contribution to the all important question of tenant-mix would be necessarily limited. It is clear that space in the high street will be let to whoever offers the best rate, but at least an organisation of this kind would be able to put forward a united argument against key properties being occupied by betting shops, estate agents, building societies, banks and offices. All of these are essential to the high street but the nature of their business so often deadens the appearance of the high street. Ideally, such businesses should be centrally located, perhaps in a court just off the main high street. They are all important to the high street as they generate pedestrian traffic.

How can all this be achieved? The local authority, with the support of the County Council, has a vital role to play. They must recognise that it is in their interests to help the high street to flourish. Successful high street trading is good for the town as well as for the local authority. More people are attracted to the town from outside, which

THE FUTURE OF THE HIGH STREET

must be good for the town generally. The local authority will be encouraged to develop the visitor potential of the town and to undertake environmental improvements beyond the high street. This, in turn, will create more jobs. Good shops and a pleasant environment attract more people; small businesses might therefore be tempted to the area. Offices and industry will follow and it will quickly become obvious that an initial attempt to improve the shopping potential of the town has spiralled healthily into something which has implications for tourism, business and industry.

The local authority should, therefore, play an active and positive role in setting up any high street association. This would bring together the local authority with representatives of the traders as well as of the shopping public. Ideally, four or five of the biggest retailers in the street should join forces to support the association and to play leading roles in its work. It would be important to ensure that this representation be drawn from senior members of staff of the appropriate calibre; managers of smaller shops seldom have an interest in matters beyond the front entrance to the shop and are not authorised to take the necessary initiatives. In choosing representatives of the shopping public, nominations could be sought from the local amenity society, the Women's Institute or Townswomen's Guild and other voluntary organisations with a direct interest in the high street. It could be helpful, too, to enlist the support of the local branch of both the architectural and the planning institutes as well as the Regional Tourist Board. The setting up of an association dedicated to improving the high street environment is so important as to suggest that the Mayor or Chairman of the Council should call the initial meeting and should be in the Chair. Other Council departments which would be represented would include planning, leisure and amenities and technical services. The association would, nevertheless, be independent of the local authority; working closely with it but allowed to express its own opinions.

Such an organisation would not be unlike the Joint Committees the Civic Trust advocated in the late 1950s and early

THE FUTURE OF THE HIGH STREET

1960s to undertake street improvement schemes. The first of these was in Magdalen Street, Norwich, where a pilot cooperative improvement scheme was launched in 1957 to 'redecorate the street, emphasise the good points, camouflage the bad and eliminate clutter'. This was followed over the next decade by scores of street improvement schemes all over the country; the proposal now being made could be regarded as building on the success of these earlier schemes by developing and extending the idea. The terms of reference of the new associations would, of course, be much broader than was the case in the 1960s. They would need to be concerned with the economic well-being of the street, its infrastructure, and the practical measures, sometimes including structural changes, required to improve the environment of the street and make it a really pleasurable place to be in. Its overall appearance and the need to maintain a high standard of display, design and lighting would naturally follow as a result of overcoming the economic and environmental problems.

The association's committee of management would need to promote itself and its aims and be provided with generous support from traders in the high street. It would be providing a service to the traders and would need to charge a membership subscription to cover not only the costs of its staff and overheads but also the cost of professional advice commissioned for any specialist commercial, economic, architectural, estate and environmental studies which the committee might consider necessary. This suggestion does not overlook the fact that there already exist a number of high street associations which undoubtedly make a step in the right direction. It is necessary, however, to broaden the base of these associations by bringing in the local authority, local organisations and the shopping public as partners. By bringing in the local authority, the committee will have the bite and authority that the street associations currently lack. It cannot be too strongly emphasised that the constructive role of the local authority is all important. They have the power to give or withhold consent at planning application stage; they have the power – many would describe it as the duty – to serve enforcement notices on those who act without consent. They also have the power to levy a rate and to assist in financing such comprehensive street schemes as would be required to provide pedestrians with protection from the elements. It is also up to them to monitor work to ensure that its quality is as high as it needs to be to satisfy the standards of the high street. If local authorities shrink from the often difficult task of upholding high standards, they would be able to quote the street association as the source of demands for quality design and workmanship.

The high street really needs the equivalent of a local doctor who will take care of architectural and environmental problems before they become obvious. It is easy to recommend, as this report does, that more use should be made of projecting signs, but the uncontrolled use of too many could create visual chaos. Advice on visual problems needs to be on hand and readily available; it does not call for a full-time appointment but when the local authority cannot provide the necessary expertise it is a service which a reliable, locally based consultant could provide. His could be the job of overseeing work in the high street, of coordinating efforts to enhance and improve and, above all, of coordinating the work of statutory undertakers, notoriously careless in the 'after-care' they provide.

The Task Ahead

Over the years standards rise and, perhaps imperceptibly, improvements in the environment do take place. The high street is one area in which the potential for improvement is enormous, but it calls for determined action by local authorities, for enlightened action by traders and for constructive participation by an interested public. Change will continue in the high street; to resist it is to waste an opportunity for profitable environmental gain.

As a nation we are confronted with the urgent need to reverse the widespread indifference to the environment we find all around us, particularly in shopping. We need to launch a national crusade aimed at making everyone more aware of their

THE FUTURE OF THE HIGH STREET

own shopping street, its potential as a trading success, as a nice place to be in, as a tourist attraction – quite simply as a credit to the town, the local authority, ourselves as well as to the world of commerce.